CW00373121

STRANGE MATES

The weird and wonderful world of animal attraction

STRANGE MATES

The weird and wonderful world of animal attraction

George Lewis

First published 2012 by
Ammonite Press
an imprint of AE Publications Ltd
166 High Street, Lewes, East Sussex, BN7 1XU, United Kingdom

ISBN 978-1-90770-839-8

A catalogue record for this book is available from the British Library.

Author: *George Lewis*
Editor: *Ian Penberthy*
Managing Editor: *Richard Wiles*
Illustration and design: *Tonwen Jones*

Colour reproduction by GMC Reprographics
Printed and bound in China by 1010 Printing International Ltd

CONTENTS

INTRODUCTION

If the creatures in this book could really speak and we were to ask them if they do the things described in the following pages, they would most likely answer "Of course. Naturally. Why do you ask?"

And that is the key to it all: what one species may regard as peculiar or disgusting, another may regard as humdrum, matter-of-fact and perfectly normal.

So there is nothing judgemental about this book: it merely relates the facts as they appear. However, it has been produced and will be read only by humans. They, too, are a species, and from their admittedly restricted purview there is no getting away from the apparent truth that the ways some animals mate are more than odd, they are passing strange.

If anyone needs reminding that there are more things in heaven and earth than are covered in sex education at school, this is the book they need.

If readers feel an "Oh yuck" forming on their lips – and they might well sometimes – they should be mindful of the adage that it takes all sorts to make a world. And that while the courting rituals of, say, the whiptail lizard (*page 118*) are way beyond what even the oddest human would ever be capable of, those of other creatures (notably some of the apes, *pages 78 and 84*) are much closer to our own behaviour than we would perhaps like to think.

The wearing of clothes and the power of speech do not stop us from behaving like the beasts of the field, the birds of the air, or even (heaven help us) some of the creatures in the sea.

LAKE DUCK

JUST A KISS

The lake duck is common throughout Argentina, Brazil, Chile and Paraguay. Its body typically measures around 17 inches (43cm) from stem to stern; the membrum virile *may be as long as 20 inches (50cm). Most birds copulate by briefly touching genital openings, an action known as a cloacal kiss.*

I'm not one to boast, but I don't need to because no other vertebrate on Earth is as well endowed as I am. Many birds don't have penises at all, but my drakehood is – or can be, in the right circumstances – the same length as my whole body. And how much longer than that it'd be if you straightened it out, removed all the corkscrews, I shudder to think. So, too, no doubt, do the lady ducks, but their opinions are neither here nor there, because when I want one of them I just lasso her with... you've guessed it.

BOWERBIRD

PAINTING & DECORATING

Like the bird-of-paradise, the bowerbird is a Passeriform. *Most of the 20 families live in Australia, but they are also found in New Guinea and on neighbouring islands. Some decorate their bowers with paint made of vegetable matter, charcoal and saliva, which they apply with leaves held in their beaks.*

Lots of birds pick the male with the most attractive plumage; some of our relatives are drawn in by fancy dancing. But we Aussies reckon you can't choose a mate until you've seen the kind of place he lives in. We check out loads of nests, but we only ever shack up where the decor's just right. What's essential is a single colour all over the floor, walls and ceiling if there is one, a consistency that eligible bachelors achieve by collecting hundreds of identically-coloured items – flowers, leaves, moss, grass, shells, feathers, stones, coins, nails, pieces of glass… you name it.

BOOBY

STUPID BIRDS?

Boobies are large seabirds widely distributed throughout the tropical regions of the Atlantic, Pacific and Indian oceans. They get their name from their tameness: the first Europeans who approached them were amazed that they made no attempt to fly away and concluded that they must be very stupid.

When we're in the mood for love, we gather in vast colonies at the water's edge and watch as the males do their amazing dance for us. First off they nod their heads and make jabbing movements with their bills, meanwhile lifting their blue webbed feet alternately off the ground: it's like stomping. In the second phase of the display, they stretch their wings out back towards their tails, crane their necks skyward and whistle at the heavens. When we've decided on one we like, we waddle over and get to know him better.

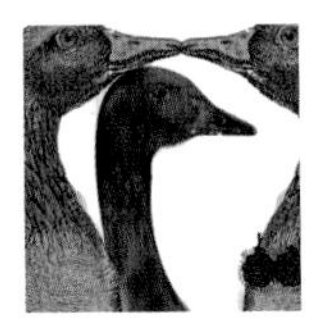

GOOSE

A GOOSE IS FOR LIFE

Most breeds of geese migrate seasonally. They travel long distances en masse in V-formations known as skeins (a group of geese on land or in the water is known as a gaggle). They mate for life, sometimes in harmonious ménages à trois *of two males and a female.*

I really couldn't care less what ganders get up to when they're on their own or think they are; I have no human-like hang-ups about same-sex relationships. In fact, they may work to my advantage: sometimes I slip between coupling males when they're too far gone to notice. And the beauty of it is that if I get pregnant as a result, they both stick around to help with rearing the goslings, so where's the harm in that? We all get what we want: this is truly the best of all possible worlds.

ADÉLIE PENGUIN

FACTS AND FIGURES

These birds are named for the wife of Jules d'Urville, who explored Antarctica in 1840. They are between 18 inches and 2 feet 6 inches (46–75cm) tall, making them medium-sized penguins: the largest, the emperors, are about 4 feet (1.2m); the smallest, the little blues, grow to no more than 16 inches (40cm).

As you might expect, given where we live, we're slow to warm up. Relationships start tentatively, with the male coyly propelling a stone towards the feet of the preferred female. (Stones are valuable to us; we use them to build our nests.) If she likes the way I roll, we stand belly to belly and sing to each other. Then it's a couple of weeks before the next move: I nuzzle my beak and my head on her breast and we take it from there. After that, we are parted for a year, but when we are reunited the following spring you should hear us holler for joy.

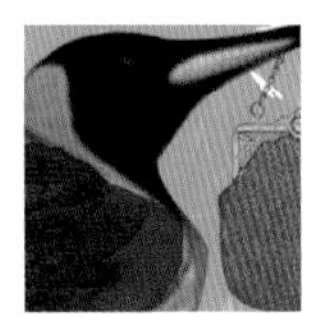

EMPEROR PENGUIN

DEEP-SEA DIVERS

Emperor penguins are slow swimmers – speeds of more than 6mph (9km/h) are the highest recorded – but they possess remarkable stamina, wandering 500 miles (800km) from the colonies to which they regularly return. They can stay underwater for up to 18 minutes while diving to depths of more than 1,600 feet (500m).

As befits birds with a noble title, ritual is important to us and our courtship is courtly: we bow low to each other before we get started. We don't have external genitalia: if we did we wouldn't be so streamlined and wouldn't move so well through the water. We each have a single multi-purpose vent, known as a cloaca, through which we pass all our excrement and reproductive material. When we mate, the female lies on her back so that the male can get on top of her and rub his cloaca against hers. It's simple, effective and anything but messy.

FRIGATE BIRD

FREQUENT FLYERS

Frigate birds are found all over the world along tropical and semi-tropical coasts. They spend nearly all their time in the air and return to their colonies only to sleep and breed. The largest frigate bird is no more than 45 inches (115cm) long, making the display ball all the more impressive.

When I'm looking for a mate, I spread my wings, waggle my head and blow up the featherless pouch on the front of my neck into a great red ball that's about the size of an adult human head. It takes a quarter of an hour to inflate. This all happens in crowded colonies where there are hundreds of other male frigate birds doing exactly the same thing. So while mating, I put my wings over the female's face to make sure she doesn't get distracted by any other bird's display: what the eyes don't see, the heart won't grieve over.

HEDGE SPARROW

CUCKOOS IN THE NEST

Hedge sparrows (aka dunnocks) are widely distributed throughout Europe and Asia, and usually live in woods, heaths, scrubland, coastal sand dunes or, most famously, in hedgerows. Although they have a reputation for promiscuity, monogamous pairs are not unknown. These birds are among the unwitting hosts of the cuckoo's eggs.

They don't call sparrows randy for nothing. During the mating season, most of us'll do it any time, anywhere with any available partner. But almost as strong as the sex drive is the need to know that all the offspring are mine. If I suspect my bird's been at it before – and it's a safe bet that she will have been – I peck at her to get rid of any sperm that may have been left behind in her behind. As I'm expected to stick around till the chicks are fledged, I want to make sure I'm not doing another bird's duty.

MANAKIN

FOREST DWELLERS

Manakins are small birds of the woodlands and forests of tropical Central and eastern South America, in a band extending from southern Mexico to northern Argentina. They are also found on Trinidad and Tobago. They are between 3 and 6 inches (7–15cm) long and weigh no more than an ounce (28g).

In the mating season, we male manakins gather in large groups, known as leks, and dance for all the females. While cavorting, we make loud noises with our throats and crack our wings together over our backs – we sound like firework displays. Some of us hold on to twigs and revolve around them like feathered Catherine-wheels, others loop the loop and do other aerobatics. When a female sees a male she likes, she joins him in the dance. Sometimes a male is such a good mover that he scores with several females at a time.

PARROT

SICK AS A...

There are more than 300 species of parrot. Although they vary greatly in appearance and behaviour, regurgitation is common to them all and can be misleading in solitary birds kept in captivity: often only a vet can determine whether the reaction is caused by desire or indigestion.

After the usual preliminaries – I parade in front of her and give her the famous eye blaze (screwing up my pupil to show her the edge of my iris) – we kiss. It starts off quite tentative and low-key, a noisy, but chaste clicking of bills. If that goes well and things start to get passionate, I experience an irresistible urge to bring up my most recent meal straight into her mouth. Some people are disgusted by this, but a female parrot likes nothing more than a bit of half-digested seed, and if she takes mine I know that I have found the right bird.

PAPER NAUTILUS

LITTLE FELLERS

*The male paper nautilus (*Argonauta*) is less than one inch (2.5cm) in diameter and virtually defenceless. The female is up to 20 times larger and has a protective coiled shell with a membrane in which her young gestate. Related to octopuses,* Argonauta *are jet-propelled and feed on plankton and shrimp.*

When humans first observed us in our natural environment – warm seas at depths between 100 and 2,000 feet (30–600m), almost never on the bottom – they thought that we were bothered by a mysterious worm, but that's no parasite, it's a detachable limb containing sperm. When I meet the right girl, I release it from my body and let it swim into an aperture on her back. Most male argonauts get only one shot at this – we're little and we don't live very long – but the mothers are big and strong and get pregnant loads of times with different partners.

BANANA SLUG

FULL OF FLAVOUR

The Pacific banana slug – a native of the western United States – is the second largest slug in the world (the biggest, Limax cinereoniger, *lives in Britain and Europe). Some people eat it – it's reputed to be particularly toothsome soaked in vinegar, though even its biggest fans admit it's an acquired taste.*

We are hermaphrodites – when I make my partner pregnant, he (or she) makes me pregnant too. Confused? This might focus your thoughts: our penises are up to 10 inches (25cm) long. That would be the envy of many creatures, but in us it's even more amazing because that's almost exactly the same length as our whole bodies, from tip to tip. The only problem with being so well endowed is that we have to be sure and find a partner of the same size, otherwise we get stuck during mating and the only way out is for the other slug to bite it off.

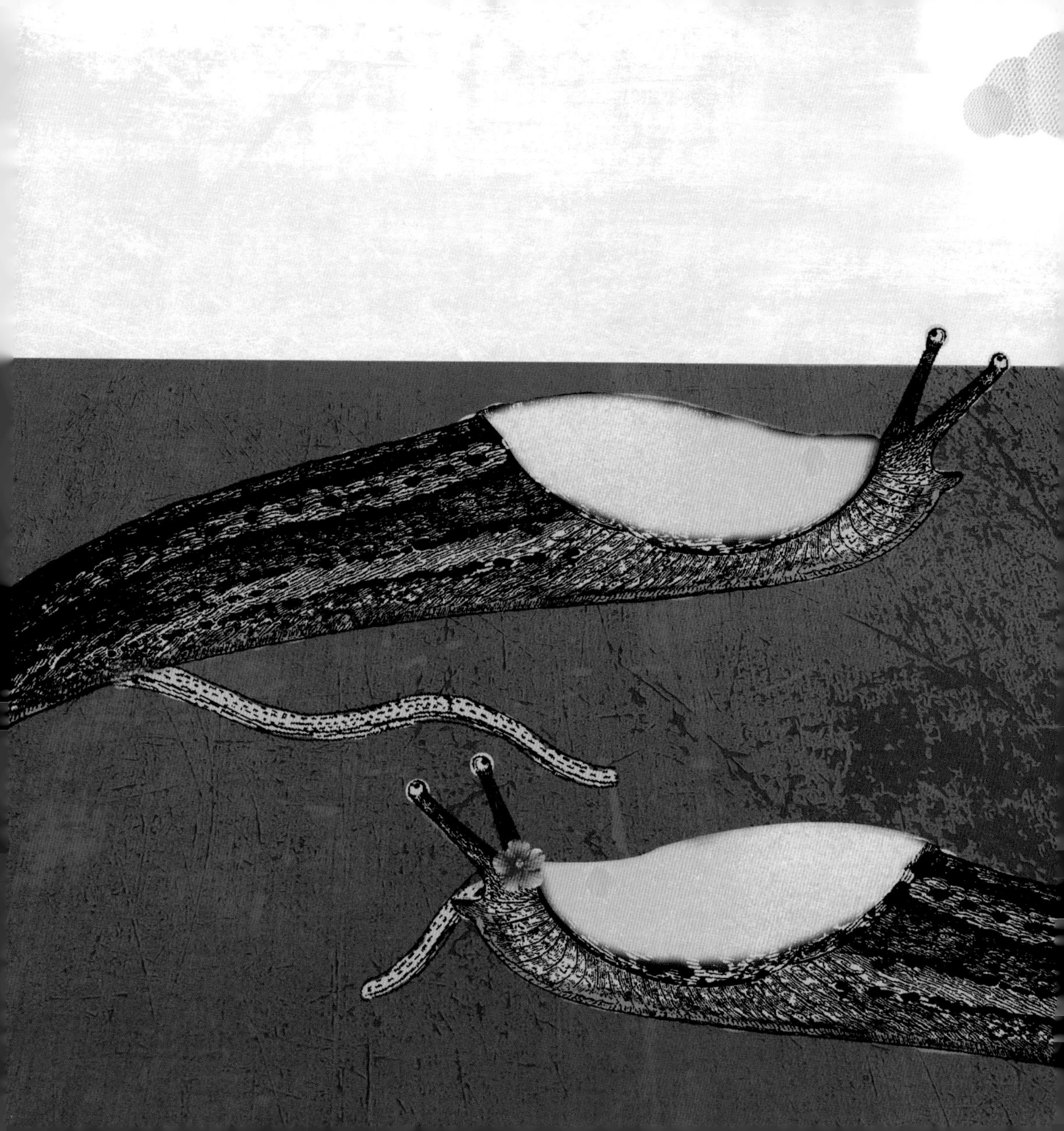

BARNACLE

HITCHHIKERS

Barnacles can attach themselves to any organism or object that spends time under water – crabs, clams, whales, seaweed, driftwood, boats. They feed by trapping passing creatures in retractable arms known as cirri. About one quarter of the more than 1,000 barnacle species are internal parasites of crabs and other crustaceans.

People often wonder how, given that I spend all my adult life attached to the hull of a ship, I ever manage to get a mate. The answer is simple: with the help of a penis that can inflate to 50 times the length of my body. I can confidently say without boasting that, relative to overall size, this is the largest reproductive organ in the natural world. And if that's still not long enough to reach another barnacle, some of us can even fertilise ourselves. Our young are born as nauplii (larvae), which swim free until they settle down and cement themselves to a likely spot.

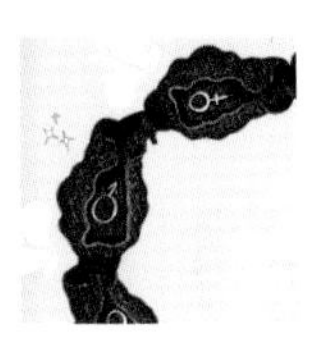

SEA HARE

EAR PRETENDERS

Aplysiomorpha *get their common name from the two long odour detectors that protrude from their heads and resemble the ears of a hare. Sea hares are up to 30 inches (75cm) long and weigh more than 4lb (2kg). They have unusually large axons (impulse transmitters) that are used in medical research into the human nervous system.*

If there's not a lot happening in the shallow waters of the Caribbean, we may have sex in what most humans regard as the 'normal' way, but since every one of us has a full set of both male and female reproductive organs, it seems a shame not to let everyone join in a game that we can all play. So when we get in a big group, we join together, boy behind girl behind boy behind girl… If you're lucky, you may see dozens of us in a complete ring, giving and receiving with equal enthusiasm.

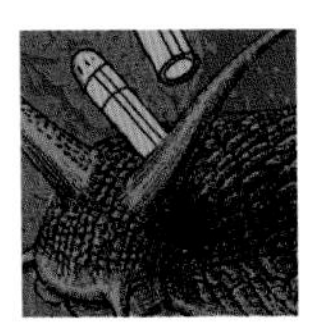

SNAIL

BEST OF BOTH WORLDS

Many young land snails have the physical attributes of both sexes, and although they usually turn out one way or the other, some remain hermaphrodite throughout their lives and thus are able to impregnate and be impregnated with equal efficiency. Mainly nocturnal, by day they retract their heads into their shells and sleep.

There are hundreds of different gastropods and they have dozens of methods of reproducing, often determined by where they live, but my way is best. When I find a mate, I shoot little calcium-cased sperm bullets at her from a launcher on my head, right behind my eye stalks. This isn't a long-range weapon; I have to get so close it may look like we're coupling, but we're not. Once she's been hit, she pouches the contents until she's ready to use them. The beauty of this is that mating can occur even when she isn't in season.

SPOONWORM

SIZE DOESN'T MATTER

Spoonworms are mainly sausage-shaped, but are named for their flattened ends, which often curve upwards or downwards at the edges. Males are so much smaller than females that until the 20th century, scientists thought they were parasites – which of course in a sense they are – but failed to realise what they really were and did.

When I give birth, I drop a load of larvae that are individually invisible to the naked human eye. At this stage, my offspring are sexless, neither male nor female. Some of them sink into the sand at the bottom of the sea, where they grow into large females like me, up to 2 feet (60cm) long. Others pick up the scent of another adult female; if they catch her, they set up home on her skin and live there until she sucks them through her nose into her genital sac, where they become males and fertilise the next generation.

ANGLERFISH

LIGHT REFLECTORS

There are more than 200 species of anglerfish widely distributed worldwide and divided into four groups: batfish, goosefish, frogfish and deep-sea anglerfish. The largest are up to 4 feet (1.4m) long. Most species are luminous, which makes their already ferocious appearance even more daunting; in Japan they are a delicacy.

My woman is big and has an amazing dorsal fin disguised as a morsel to lure her prey. When a victim comes within reach, she swallows it whole. And I live off her. Which is a cushy number until cupboard love turns into the real thing. I wiggle up and bite her – not aggressively, like a shark, but sensuously, because that's how I make her pregnant. And that's where it goes wrong because once I've done that I'm stuck to her. Forever. She draws out my vitals to feed herself and her young. So by becoming a father, I sign my own death warrant.

CICHLID

A MOUTHFUL

Cichlids are freshwater fish of tropical Africa, America and Asia. Their eggs take around two weeks to hatch, and the mother will typically keep the young in her mouth for a further fortnight, during which time she may occasionally open her jaw to give her offspring a glimpse of freedom, but she gulps them back in at any sign of danger.

When a male is attracted to me, he shakes his fins in an underwater dance. If I'm interested, I'll let him show me his nest, which he's dug out specially. Assuming all goes well, I'll lay my eggs, sometimes as many as thirty. After that, the sticks around, which isn't as good as you might think because he doesn't help with the incubation: all he wants to do with the eggs is eat them. So I have to keep them all in my mouth until they're hatched and old enough to see him off if he tries anything.

OCTOPUS

THE EGGS FACTOR

The modified arm of the male octopus is known as the hectocotylus; the seed packet is the spermatophore. Throughout a gestation period of between four and eight weeks, the female guards the eggs, cleans them regularly with her suckers and wafts the water around them to keep them at the right temperature.

You might think that the male octopus is an impressively virile creature – check out his eight strong tentacles, including one that is specially adapted so that it can impregnate me with packets of sperm, which I can use anytime. But the fact of the matter is that shortly after he's done his duty, he dies. No one quite knows why: maybe it takes more out of him than he thinks it will; perhaps he decides that his work on Earth is done. Whatever the reason, I end up holding the babies, sometimes as many as 100,000 of them.

PERCULA CLOWNFISH

THE RIGHT PARTNERS

Clownfish are found in coral reefs of the Indian and Pacific oceans, where they shelter in and around sea anemones. Their bright colours make them popular tropical aquarium fish, but they are not easy to maintain in captivity because there are only certain species of anemone with which they can co-exist.

Stars of the computer-animated movie *Finding Nemo*, we live in nuclear families: a large female and a little male, who mate with each other and breed, and a few (often two) even smaller fish who appear to be male, but actually are sexless because their gonads are inactive. The reason these hangers-on stick around does not become apparent until the female dies, whereupon one of them changes into a fully-fledged female and pairs with one of the other non-breeders, who obligingly becomes a male. What Disney/Pixar didn't tell you.

SQUID

MAKING THE WATER MOVE

Squids are found in all the world's oceans. The smallest are less than three-quarters of an inch (20mm) in diameter, while the largest measure more than 65 feet (20m) across. They congregate in vast numbers to mate, and the fighting as they pair off can make the waters churn for hours on end.

Male squids can do extreme violence when the occasion demands – with enemies and prey and rival suitors – but we've got nothing on the females, who never seem to like our attempts to mate with them. Luckily, we have the means to keep out of their grasp while we do what we have to do – Nature has given us penises that are only slightly shorter than our tentacles. Even that doesn't guarantee our safety – many's the amorous squid that has had to duck and dive so much to avoid being thrashed and bitten to death that he has ended up injecting the sperm into himself.

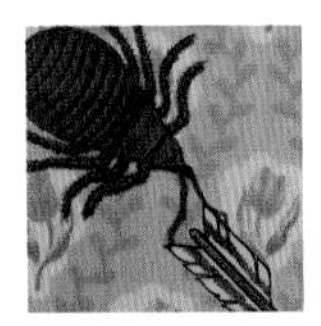

BEDBUG

BLOOD LUST

Bedbugs feed on the blood of mammals; they're especially partial to humans. They are highly resilient and hard to get rid of: starving them out is not an option, because they can survive for years without food. Their bite is maddening, but as far as is known, they do not transmit disease.

The job of impregnating a female bedbug is not made easier by her lack of an external passage to the reproductive organs. The only way to get around this problem is by stabbing her in the abdomen with my hypodermic penis. Biologists call this traumatic insemination, and while it's true that it does make a wound that takes some time to heal and may of course become infected, the damage can't be all that bad because if it was she would not then be able to lay and incubate an average of 200 eggs at a time.

EARWIG

NIGHT CRAWLERS

Earwigs also have two wings, but they are much less useful than their penises: these insects can hardly fly at all. They are nocturnal and live on fruit. The idea that they crawl into the ears of sleeping people, though widely repeated, has no basis in fact.

We have two penises. For years that was our big secret, but in the 20th century some nosy Japanese scientists tried to separate a pair of us while we were mating (why would they?). Horrified when they found that the penis the male was using had broken off, they then looked on in astonishment as he just switched to the other one and carried on with the next available earwig as if nothing had happened. Why did we get this double blessing? Simple: so that we could ensure that once we'd deposited our seed, it stayed firmly locked in place.

1
2

FIRE ANT

LONG-DISTANCE TRAVELLER

The fire ant originated in Central and South America, but emigrated in the 1930s by boat to the United States, where it rapidly acquired a worse reputation than any of its native cousins because of its nasty sting. It has since made it to Australia and China, where it's every bit as unpopular.

I'm the queen, the biggest ant in the hill. I have millions of subjects and they are all either drones (males that live solely to fertilise me and die as soon as they have done so) or workers (sterile females that do all the practical stuff around the nest). It's important to keep the numerical balance between the two classes just right, so whenever we get too many of one lot I get the other lot to kill them while they're young (if you leave it too late, it gets nasty, because adults of both sexes are mighty aggressive).

FRUIT FLY

ILLEGAL IMMIGRANTS

These two-winged insects lay their eggs – up to 500 at a time – in citrus crops. The larvae then burrow into the fruit, making it unfit for human consumption. They originated in the Mediterranean, but are excellent travellers, which is why strict international quarantine laws regulate the export and import of fruit.

The reproductive tract of our females is twisted up like a labyrinth and full of chemicals that destroy all but the most formidable liquids inserted into it. This might have threatened the survival of the species had we males not been endowed with vast sperm that, if straightened out, each measure 2 inches (5cm), more than 1,000 times longer than their human equivalent. This is pretty impressive when you think that our whole bodies are no more than a tenth of an inch (2.5mm) long. Even more amazing is that our testes make up more than one tenth of our total body weight.

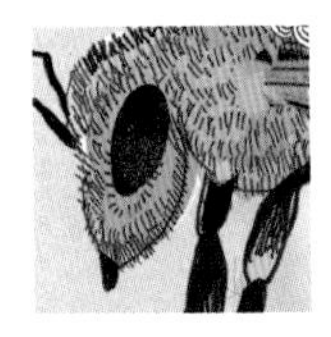

HONEY BEE

PLANT PROPAGATORS

Bees feed only on pollen and nectar, and store some of the latter in the form of honey. As they fly around collecting pollen, they often deposit some of the fertilising powder of one flower on another. Thus they are vitally important agents of pollination.

My existence is clearly mapped out from start to finish: I live a life of ease; the workers (poor, unenviable cousins) bring me all the food and drink I need, and I do nothing except prepare myself for one magnificent, defining act of sex – not just random sex, but sex with the queen bee herself. The downside is that as soon as I've had my way with her, my genitals explode to seal the entrance to her womb. Naturally I die straight away. Some people pity me, but the way I look at it, no pain no gain; and what a way to go.

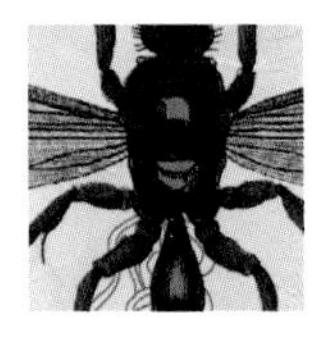

ICHNEUMON

MANY IN NUMBER

The 4,000 species of ichneumon are related to and may resemble wasps, though they vary a lot in appearance and especially size: the smallest are around half an inch (1.2cm) from end to end; the largest genus, Megarhyssa, *natives of North America, may be four times as long.*

We may make our home in the young of several species, including beetles, moths and sometimes even spiders, but we're particularly fond of caterpillars. Wherever we end up, we inject the accommodating hosts with our eggs and a poison that paralyses them. As soon as they become helpless, we start eating their fat. We keep each victim alive for as long as possible, saving its heart and other vital organs till our young are ready to enter the world. Humans don't normally like parasites, but we're a special case because we get rid of many insects they regard as pests.

PRAYING MANTIS

IT'S A MYSTERY

It's not entirely clear why these insects are described as 'praying', although their name has religious overtones in many languages and dialects – in English 'mantis' means diviner or soothsayer. Given that they are all ferocious carnivores, the name looks like a mis-spelling of 'preying'.

There's no denying that we females have been known to bite the heads off our mates as soon as they've finished copulating. But there's a good reason for that: it doubles the chance of fertilisation. Why that should be the case is a whole other question that even we don't understand: it may be that a frightened male is more potent; or, perhaps females that are hungry, and therefore likely to eat their mates, are more fertile than those that are already full up. But nothing is a dish for every day: we really can and often do have sex without killing our lovers afterwards.

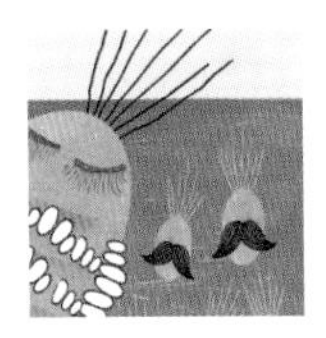

HISTIOSTOMA MURCHIEI

HARDY TYPES

Most mites are only around one two-hundredth of an inch (0.1mm) long, so are invisible to the naked eye, although some grow to as much as a quarter of an inch (6mm). They are awesomely adaptable and can thrive in almost any environment or, in the case of parasitic species, on any host.

They call me a parasite. And while it's true that I live on the cocoon of an earthworm, I don't like the word because it has bad connotations. It's a word that, once applied to something, makes people think that's all they need to know about it. And then they'd never find out that I lay two kinds of eggs. The first is a small clutch that turn into males without being fertilised. Then I mate with my newly created companions to produce a batch of around 500 more eggs that produce females like me. They should call me the life-giver.

RED VELVET MITE

COLOURFUL CHARACTERS

Red velvet mites are widely distributed, but most common in India. They are named for their bright colour and sleek appearance; the latter derives from the fine hairs that cover their backs. They are carnivorous, feeding mainly on the pre-digested flesh of insects. Although they have eight legs, they are not spiders.

When I need a mate, I deposit my sperm on a small twig or a plant stalk and then spin a line of silk that leads away from the spot and out into plain view. If a female sees this thread and likes the look of it, she will follow it along all its twists and turns – a scenic route is a nice touch – and then impregnate herself by sitting on the package I've left for her. Once I've laid the trail, I need to keep my eye on it in case another male comes along and tears it up.

STRAW ITCH MITE

IRRITATING BEHAVIOUR

Straw itch mites give birth to between 200 and 300 young at a time. They're especially fond of living on humans, in whom they can cause a nasty red rash. Although most attacks are more irritating than harmful, some doctors identify the straw itch mite as a cause of eczema.

We have a lot to cram into a short time – we usually live for only about a week, almost never for more than a fortnight. So as soon as we emerge fully formed from our mothers, we hang around by the exit we've just come through and wait for our sisters to show (they're always a bit later than us males). And as soon as they land, we get straight on and mate with them. Pretty soon after that we all get blown away by the wind and settle on new hosts, and the cycle begins again.

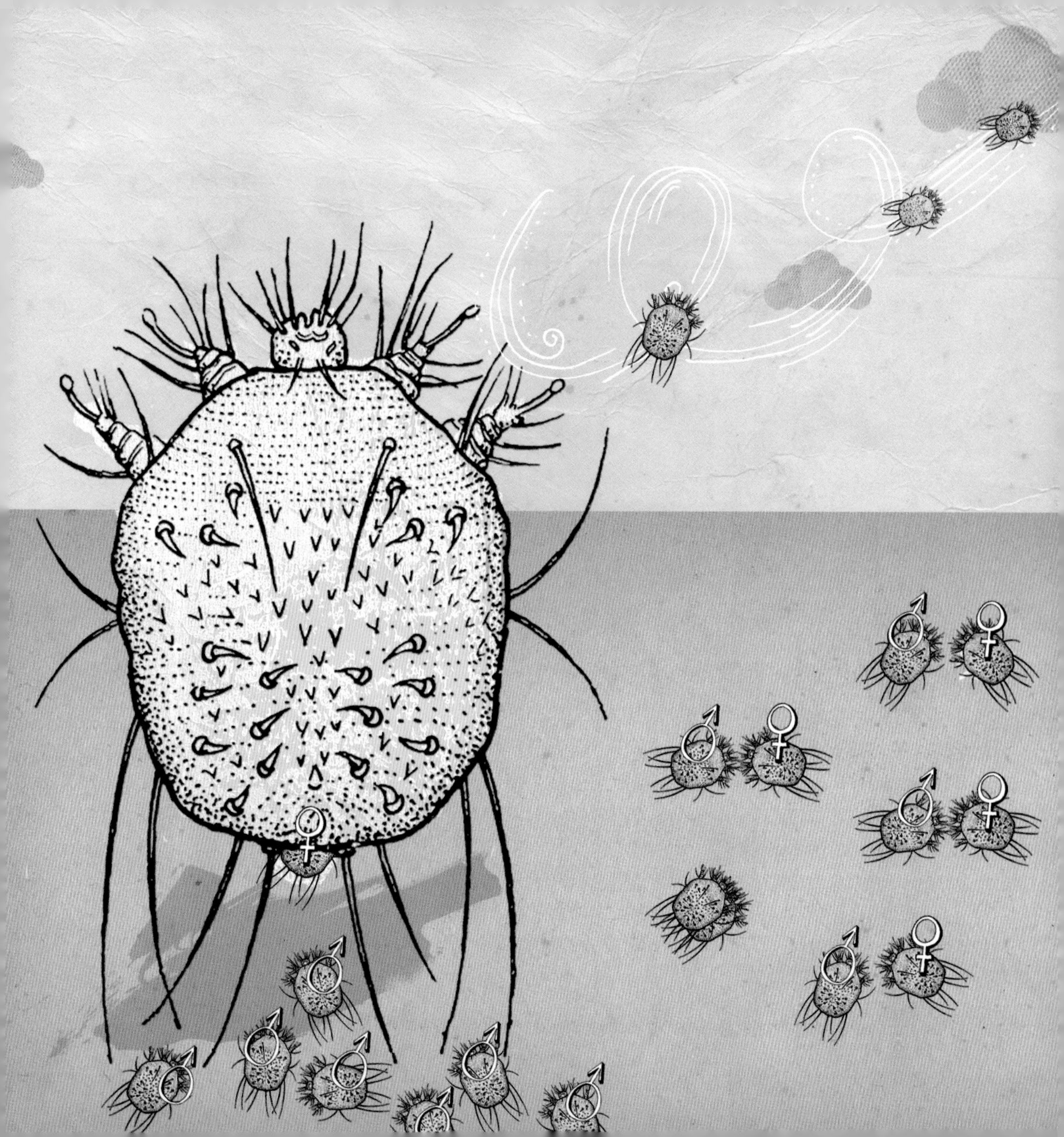

WATER STRIDER

DANGEROUS GAME

The water strider's main predator is the backswimmer, an insect that walks upside down immediately below the surface of the water. Thus when a male water strider climbs on a female's back, she is in a highly vulnerable position for the duration of the mating.

We're those insects you see hopping around on the surface of rivers and ponds. Some people call us Jesus bugs. Our females have genital shields that enable them to refuse all but the most eligible males. And they're very choosy: if they had their way, no suitor would be suitable enough. But fortunately they don't have it their way: we males have developed a cunning strategy; we give off vibrations that attract not only females, but also our main predators. So, being smart, the females take the line of least resistance and let us get it over with asap.

SCORPION

SINGLE PARENTS

Scorpions vary widely in length, from around half an inch (12mm) in the Caribbean to 8 inches (20 m) in South Africa. Not every species mates in the chilling way described: some reproduce without the involvement of a partner, a process known as parthenogenesis.

I am drawn by her irresistible scent. When I get close, I grab hold of her with my front pincers and we shimmy sideways in what non-scorpions may liken to a dance, but in fact is an attempt to find a suitably smooth surface on which to mate. Once we've hit terra firma, we get on and do what Nature intended. That's when the real excitement begins, at least for the males, because then we have to get off them before they can bash us with their stings: those who fail to pull away in time get eaten.

FLATWORM

LIVING OFF OTHERS

Some flatworms live independently in tropical fresh water or sea water, but more than three-quarters of the 20,000 species are parasites within the body of a host. Although generally regarded as a pest, they have been used in several parts of the Pacific Rim to control the imported giant African snail.

We are hermaphrodites – every one of us has ovaries with eggs and testes with sperm. Although we're perfectly capable of reproducing without a mate, it's normally preferable to do it in partnership. But the problem is that no flatworm wants to do the work of the female, caring for the eggs, so we fight for the right to take the male role. Our weapons of choice (indeed, our only weapons) are our penises, with which we fight like swordsmen to assert our dominance. The winner is the first worm to pierce the body of the other and impregnate it.

BROWN ANTECHINUS

ONCE IN A LIFETIME

The brown antechinus lives in the rain forests of Australia, Tasmania and New Guinea. It has a hairless tail like a rat, but bigger eyes than the Eurasian rodent. Antechinuses eat invertebrates such as spiders, beetles (including larvae) and weevils. Creatures that mate only once in a lifetime are said to be semelparous.

We're marsupials – Australasian mammals that carry their young in pouches because they're born too weak to survive any other way. When we get to the age of about 10 months, we gather together in a big group and spend more than half a day mating with as many partners as we can grab. Then the females give birth to the offspring of several males – typically three or four different fathers – in the same litter. The effort of all that sexual activity causes a complete and rapid breakdown of our immune systems; we seldom make it to our first anniversaries.

BONOBO

NICE TO KNOW

Almost exclusively herbivorous, bonobos live in rainforest trees along the banks of the Congo River in the Democratic Republic of the Congo in Africa. They were known as pygmy chimpanzees until the 1930s, since when they have been classified separately. No one knows why they're so pacific: it may be because their environment provides their every need.

We're related to chimpanzees (though not as closely as humans once thought), but we are much more peaceful than our disputatious cousins, who fight and kill each other over very little. Which is not to say that there are no rivalries – we have rows about territory like any other mammal, and our males often try to outdo each other – it's just that we settle our quarrels by making love not war. We do a lot of fondling and kissing, and when we have sex we often do it face to face, which again unusual is in the animal kingdom.

CALIFORNIA SEA LION

CLEVER ANTICS

California sea lions live along the west coast of the United States and Mexico and in various parts of the Pacific as far as Japan. Highly intelligent, they're the breed that performs in circuses and zoos. One of the mysteries of science is why they are never found in the Atlantic Ocean.

At the start of the mating season, we bulls stop eating and set ourselves up on a prominent spot where we can see and be seen. Once we're ready, we bark at the cows to attract their attention. And then they come milling past – they're never alone, they cruise in pairs or groups of as many as 20. If they like what they see, they come in for a closer look and might even hop on top of us for a sort of test ride. As soon as they've made up their minds, the milling stops and the real action begins.

CAT

SPEAKING OUR LANGUAGE

Cats make noises to communicate with humans that they never make among themselves. Their purring is commonly thought of as an expression of contentment, but they have been heard to do it when they're in pain. Some scientists think purring is a sort of feline mantra.

Cat mating is a noisy business. When we're looking for mates, we cry out to each other – it's called caterwauling – and after we've finished with each other, the tom makes the queen scream as he withdraws because he has a ridge of about 150 tiny, but sharp backward-pointing barbs on his penis. Some people think that's just plain cruel, but there's a purpose to it all: ripping the vaginal wall stimulates ovulation. Besides, the she-cats aren't seriously hurt. And as with so many other cat behaviours, the same phenomenon can be observed in all species, from the domestic moggy to the Siberian tiger.

CHIMPANZEE

PLAYING AWAY

Male chimpanzees tend to spend their whole lives in the community in which they were born and, though seldom solitary, often seem content on their own for long periods. Females are more sociable, and more likely to roam and mate with males from other groups, thus reducing the danger of inbreeding.

Humans who watch us have observed that younger female members of our community may offer to engage in acts of intimacy in return for payment in fruit. They have also noticed that both sexes are what they regard as highly promiscuous and likely to get fighting mad through jealousy. Moralists disapprove of these practices, but we live in a highly competitive world where alpha males bully and argue their way to the top of the pile: we know what we want and we take it if we can get it, behaviour that perhaps is not unlike that of humans themselves.

DOLPHIN

MISTAKEN IDENTITY

There are 32 species of dolphin. Most are only around 10 feet (3m) long, but the largest – known confusingly as the killer whale and the pilot whale – may be over 26 feet (8m). The most popular is the bottlenose dolphin, the one with the mouth curved in a fixed smile.

For thousands of years, dolphins were thought of as models of true love and monogamy – in religious symbolism that's what we always represent. It is only in the last hundred years that we've been unmasked for what we really are: some of the randiest creatures on Earth. We will try to do it with anything – turtles, rocks, driftwood, people… This is not completely surprising when you consider the adult dolphin penis, which is not only incredibly sensitive, but also prehensile: we use it to feel and grasp things in much the same way as humans use their hands.

ECHIDNA

PRICKLY DOWN UNDER

Also known as the spiny anteater, the echidna is the most common native animal in Australia and the second most common creature of all after the house mouse, which was introduced from Europe. Echidnas are monotremes – animals that excrete and reproduce through a single bodily orifice.

We're one of only two egg-laying mammal species in the world; the other is the duck-billed platypus. There is an old joke about how we need to be careful when making love because of our spikes, but they're not an issue; we're used to them. Our main problem is that there are too few female echidnas to go around, so we males have to take part in orgies where there are lots of us and only one of them. Sometimes it becomes so desperate that we're forced to cut short our winter sleep and mount she-echidnas while they are still hibernating.

GIRAFFE

GANG CULTURE

Giraffes are not particularly social animals, although males (bulls) hang out together when they are young. At this time, fights are common, and although there is no serious intent many adults are scarred for life by playful head-butting. In the mating season, bulls may travel for days in search of likely females (cows).

When I choose a mate, I use the tried-and-trusted method of prodding her up the backside to make her pee. Then I take a mouthful of her urine, the taste of which will tell me if she's in season. If she isn't, I move on: there's no point in going to all that trouble if there's no chance of a calf. If she is, I get on with it at the first possible opportunity, although that may take a few hours to present itself because, before yielding, the female always likes to make sure there are no more attractive males around.

GOAT

HUNGRY FOR SEX

There are 69 breeds of goat. They reach puberty at between 3 and 15 months. During the mating season, both sexes lose their appetites and the females stop producing milk, no doubt so that they can devote all their time and energy to rutting. The tail-wagging is known as flagging.

We're always at it. Kids only a week old go through the motions with anything they can mount – sex and species do not concern them. And that goes on into adulthood. Nanny goats show their availability by frantically wagging their tails at the billys. The males then pee all over themselves – face, beard, legs – and approach the she-goat, who squats and does the same so that he can sample her urine. If it's to his taste, the couple walk around side by side for a while making whooping noises that are the prelude to the main action.

GORILLA

NOT SO WELL ENDOWED

Gorillas live in troops of up to 30 in West Africa and Eastern Central Africa. Adult males may be five and a half feet (1.7m) tall and weigh as much as 500 pounds (227kg). Their penises are typically no more than one and a half inches (4cm) in length.

Around 99 per cent of our DNA is the same as that of humans, and many aspects of our behaviour are similar, too. Take sex: male or female may take the lead, depending on circumstance. If it's her, she first establishes eye contact, pursing her lips provocatively and walking slowly towards the target. If he ignores her, she'll attract his attention by slapping the ground with her knuckles. If it's me, I'll show her my genitals which, though not the largest in the animal kingdom, relatively or absolutely, impress her well enough and are perfectly adequate for purpose: size really doesn't matter.

HIPPOPOTAMUS

SNEAKY TYPES

Hippo are regarded as Africa's most dangerous wild animal: they sneak up unnoticed on land or in water and kill if they can. Their skin often appears to be covered with red fluid, which gave rise to the legend that they sweat blood; in truth, the colour comes from a natural secretion that acts as a sunblock.

Apart from obvious differences in size, shape and swimming prowess, what mainly distinguishes a bull hippo like me from human males is that while they normally do their business in private, mine can be a major attraction, especially during the mating season, which starts every year after the rains. There's nothing cows like more than if I perform my evacuations in front of them, whirling my little tail like a propellor so that the solids and liquid get sprayed in a great arc all over the place. The one that chooses me then leads the way down to the water…

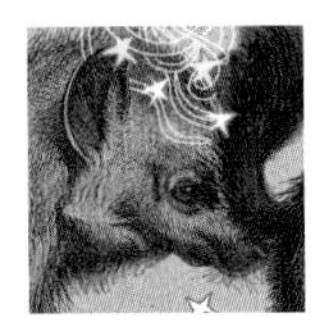

HYENA

DOGS OR CATS?

Hyenas look like dogs, but are more closely related to cats. They have bad reputations as scavengers and carrion eaters, but they're efficient hunters, too: they have been known to kill anything from fish to hippos and zebra. They live in packs of up to 80 in many parts of Africa and Asia.

Externally, the genitals of hyenas of both sexes are indistinguishable. The only difference is that while the females can produce erections at will, the males can do it only in response to the normal stimuli. Part of the fun is working out by close inspection and sniffing which is a pseudo-penis and which is the real McCoy. It's not without risk, though, because we're all aggressive and equipped with powerful jaws that we're not afraid to use. Some of us do end up with our manhoods mangled, but not so many that it endangers the species.

MACAQUE

THEY'LL EAT ANYTHING

Macaques – the monkeys in the Buddhist saying 'hear no evil, see no evil, speak no evil' – live mainly in Asia, but one species, the Barbary macaque, is native to North Africa. They are omnivorous and although they seem to like trees best, they are equally at home on the forest floor.

The way to win over a female of my species is by grooming: the longer I comb her hair, the more she will let me have sex with her, sometimes as often as four times an hour. The latter is a noisy business: it doesn't work for us unless we both cry out at the moment of climax. And the more we do it, the less likely she is to run off with another macaque. Which may be the root of one of our biggest social problems: solitary macaques creeping up on mating couples and attacking the male from behind.

MOOSE

DANGER MOOSE

Although moose (aka elk) are commonly associated with Canada and Alaska, they are also widely distributed throughout Siberia. Russians are particularly wary of them, regarding a wounded adult (and many of them are seriously injured by love rivals during the mating season) as scarier than a bear.

Autumn starts early in the far north – late August – and as soon as it does, we start rutting so that the calves will be born the following spring. If you knew no better, you might think that the males make all the running, but the females are sly: they leave their attractive pee in places where they know we'll find it. And once we've tracked them down, the sneakiest of them give off piteous moaning noises that make them sound like they're in trouble. Other bulls come rushing to the rescue and start a fight; the scheming cow watches from the sidelines and then goes off with the winner.

PANDA

PANDA PORN

In the 1960s, London's Chi Chi and Moscow's An An were brought together twice, but wouldn't go near each other; the only sexual interest the former ever displayed was in one of her keepers. More recently, zoos have tried pandas with Viagra and shown them blue movies of other pandas having sex: neither made much difference.

We're one of the ten most endangered species in the world, not so much because of persecution or threats to our habitat as because the females are in season for no more than three days a year and the males don't always feel like obliging them at the critical moment. In the wild, most of the time, even if we wanted sex, we wouldn't be able to have it because we're solitary creatures who usually live many miles apart. In captivity, we go off sex completely, at least with members of our own species.

PORCUPINE

DOING IT IN THE DARK

Porcupines look like hedgehogs and echidnas, but they're related to neither species. They are mainly nocturnal and live in a wide range of topographies in tropical and temperate parts of Asia, Southern Europe, Africa, and North and South America. Their quills or spines are modified hairs made of keratin.

At the start of the mating season, we break off from our normal routine and start rubbing our genitals against convenient objects, peeing on them if the spirit moves us. As the build-up intensifies, we concentrate our efforts on places where members of the opposite sex have rubbed themselves or peed before. When we finally choose a mate, we rear up and stand on our hind legs, supporting each other with our forepaws. Just before we go belly to belly, I pay the female the ultimate compliment: I hose her from head to tail with a magnificent jet of urine.

UGANDAN KOB

A PATCH TO CALL THEIR OWN

These increasingly rare African antelope resemble impala, but are smaller and more thickset. The males' individual breeding territories are remarkably small – no more than 100 feet (30m) across, sometimes only half that size. There are between 30 and 40 adult males in each herd and at least as many females.

At the start of the mating season, we each mark out a little territory of our own and wait for the females to come looking for us. It's exciting when they first appear, but we should beware lest our dreams come true. They all have identical sexual preferences, so they all go for one male, and he has to service them in turn. Or try to. Naturally he can't manage, so when he's exhausted, the herd moves on to the second-choice male, and so on. They give a whole new bad meaning to the slogan 'All for one and one for all'.

GALÁPAGOS TORTOISE

A RIPE OLD AGE

*Galápagos tortoises are native to the Pacific islands of the same name (*galápago *is Spanish for tortoise). The largest weigh 880 pounds (400kg) and are nearly 6 feet (1.8m) in length. Their lifespan is the longest of all vertebrates: typically they reach their centenaries, and one individual is known to have lived for 170 years.*

The first thing to do is see off the competition. Male rivals go toe to toe and stand on their hind legs, stretching their necks and opening their mouths as wide as possible. This may be the prelude to a fight, but normally the smaller giant knows not to try to punch above his weight. The foreplay's not without aggression, either: we butt the female's shell and sometimes nip her legs. The biggest problem of all is the actual mounting: the effort of stretching and holding myself on top of her shell makes me grunt loudly.

Aaaaaagh!

KING COBRA

SHORT-TERM RELATIONSHIP

The world's longest venomous snakes, king cobras may be up to 18 feet (5.6m) long. They live in the forests of India, Southeast Asia, Indonesia and the Philippines. They build nests and couples stay together after mating until their eggs are hatched, between 60 and 90 days later.

In the breeding season, we all shed our skins and the females give off pheromones that lead us to them. If I see a likely candidate before she sees me, I go straight in and coil myself around her; otherwise, we rear up and sway in front of each other. We have to be careful though: things might turn nasty. Even if she lets me snuggle up, she may still try to slide away, but I hang on until she gives in to the inevitable. Then we remain in each other's coils for several hours while I locate her cloaca and deposit my sperm.

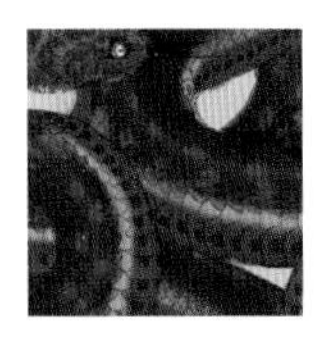

RED-SIDED GARTER SNAKE

NOTHING TO BE SCARED OF

Garter snakes are non-venomous serpents of North and Central America, where they are thought to be the most widely distributed reptile. Adults are around 39 inches (1m) long. Their main diet consists of insects, earthworms and small amphibians. They do not lay eggs, but give birth to live young.

We hibernate en masse, in groups of as many as 30,000. When we wake up again in spring, the first thing we think of isn't food, but sex. As there are so many of us coiled all around each other, we just start going through the motions regardless of whether the nearest garter snake is male or female. Most of us expend a lot of effort for no return, but some of us get lucky and find a suitable partner, although some males give off pheromones that make others think they've scored with a female when they haven't.

RHEOBATRACHUS FROG

NO MORE LEFT?

This gastric-breeding frog of Queensland, Australia, has not been seen since the 1980s and some naturalists fear that it may be extinct. The chemical in the jelly is prostaglandin, a complex organic fatty compound. Young rheobatrachi are born fully formed: they go through the tadpole stage in the mother's stomach.

Once my eggs have been fertilised by a male, I swallow them. It may look like cannibalism, but fortunately my future offspring are surrounded by chemical gel, which neutralises the acids that would normally break down anything in my stomach. The system works well in general, but there are two drawbacks: one is that I can't eat anything throughout my pregnancy, six long weeks; the other is that the number of young I eventually spawn is never more than about half the number I ingested, though that doesn't necessarily mean I ate them; it could be that I dropped them on the way in.

WHIPTAIL LIZARD

OFFICIAL RECOGNITION

The mock mating ritual performed by whiptails has given rise to the nickname 'lesbian lizards'. Up to four eggs are laid, hatching around eight weeks later. The whiptail is the official state reptile of New Mexico.

We live in the desert grasslands of Arizona, Texas and Mexico. The eggs we produce develop unaided into fully-formed lizards in the exact likeness of their mothers. The process occurs without fertilisation by males, which is just as well because there are no male whiptail lizards: they simply don't exist because there's no need for them. Reproduction is not entirely a solo effort, however: it takes another female lizard to start the whole process off by mounting us and simulating sexual intercourse: scientists call this pseudo-copulation; ordinary people have a ruder term for it: dry humping.

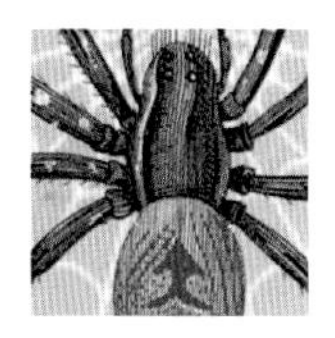

BLACK WIDOW SPIDER

THEY'RE EVERYWHERE

Black widow spiders – sometimes known as brown widows or redbacks – are found on every continent except Antarctica. The bodies of adult females are about one inch (2.5cm) long; their outstretched legs may more than double their overall size. Their main prey is insects, and their main predator wasps.

It's quite untrue that we always kill and eat our partners after mating with them. But we often do. There seems no reason not to: they're small – a quarter of my size – and rather insignificant, and after they've impregnated us they have fulfilled their eugenic purpose. They're not even very good at being poisonous: while we can do humans a power of no good, killing the very young and the sick if we get our teeth into them, they hardly notice our menfolk's bite. And beware of false widow spiders: impostors that have our markings, but aren't as venomous as they try to look.

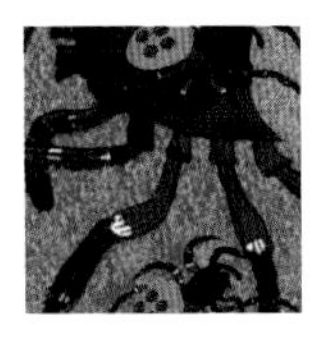

LADYBIRD SPIDER

MATING MARKS

This spider is named for the red and black markings that the male of the species develops on his back during the mating season. For many years, it was one of the rarest species in England, but careful protection has helped its numbers to revive, particularly in Dorset.

We're endangered, all right, not just by urban sprawl and pollution, but by our own nature: we give up our lives for our young ones. The male puts his all into the act of mating: so much so that he dies almost immediately he's done it. The female then devotes all her energy to nurturing her eggs, moving the sac in which they develop around her underground burrow at least twice a day to make sure it is always at the right temperature. And when the young ones are born, they hang around in the nest until their mother dies, whereupon they eat her.

RIP

LYNX SPIDER

LIKE A CHAMELEON

Lynx spiders do not build nests or conventional webs, but capture their prey by pouncing on them in a cat-like way – it is from this that their name is taken. They can change their coloration to match that of the plant or flower on which they are lying in wait.

When I find the female I want, I wrap her up tightly in a silken web so that she can't get away. Then I go off and catch loads and loads of insects. When I think there are enough of them, I unwrap the she-lynx and invite her to eat as much as she wants. Which she does. And while she's feasting, I creep up and mount her from behind: the way to a female's heart is through her stomach. You can see us in action in almost every part of the world apart from the Arctic and the Sahara and Arabian deserts.

TARANTULA

A LIKELY STORY

This hairy spider of the Western Hemisphere measures up to 3 inches (7.5cm) in diameter. The popular belief that its bite causes tarantism – a disease that makes victims dance around madly in their death throes – is entirely without scientific foundation. Tarantulas may deliver a nasty nip, but they are not poisonous to humans.

First I climb all over a specially prepared web, rubbing it with sperm, which I pick up in my pedipalps. When I find a suitable female, I twitch my abdomen and drum my legs as I approach. If I'm in luck, she may make reciprocal gestures. Once I get within reach, I grab her fangs with my mating hooks, lift her up and make my deposit. No matter how long this takes – it could be seconds, it could be hours – as soon as it's over, I leg it fast: females of my species can be every bit as deadly as black widows.

AMMONITE
PRESS